My
Koi Keeping
Book

A Guide for Koi Keepers

By Susan Trott

Copyright

My Koi Keeping Book (Second Edition) blends practical wisdom with record-keeping pages, checklists, and conversion charts. Drawing on nearly thirty years of experience, Susan Trott offers koi enthusiasts an inviting space to record, reflect, and care for their fish year-round.

By Susan Trott
ISBN 978-1-998107-49-0
Printed in the United States by IngramSpark

Tagger Press

Dedication

This book is dedicated to all the people who have helped us get to this point, but most importantly, my husband Jim, who when I said "I think I'll build a pond in the backyard" didn't blink. Instead he's been my partner in our nearly 30-year adventure keeping Koi.

I'd also like to mention good friends who have helped, Gavin, Helder, and Gary. Their wisdom, assistance, and knowledge have been instrumental to us. We learned so much, and our fish thank you.

And to our Koi. They've thrived in spite of us.

Contents

Introduction

This guidebook was created with the first-time fish keeper in mind. It offers a clear starting point for good animal husbandry and the rewarding, sometimes epic, journey of koi keeping. At the same time, it will be useful for long-time hobbyists who want to refine their practices or bring more structure to their care routines.

The Journaling pages included here grew out of my own experience. Over the years, I found that no matter how dedicated I was, life has a way of interfering—and important details can slip from memory. Having a place to log water tests, treatments, and observations quickly became an indispensable tool.

Recording information formally does more than preserve memories. It helps reveal patterns and highlight problems early—often before they turn into disasters. A simple habit of logging can save you and your fish a great deal of stress.

This is the Second Edition of the Koi Keeper's Journal, expanded and improved since the first version two years ago. Along with the log pages, you'll also find appendices with conversion charts, plumbing references, and adaptable checklists—resources you can tailor to your own pond and style of keeping.

My Koi Keeping Philosophy

There are many different approaches to keeping koi but there are two predominant ways:

1. A koi-only pond, with bottom drains and a bare liner
2. A natural looking pond with koi, rocks, and plants

I personally ascribe to the first one, because we've been through the route of second one already.

Now, I'm not going to preach. I don't really have any 100% argument for either way, it's what you can afford, what you like, and what you want to build.

However, that said, there are a couple of things that people from the second category can and should adopt from the first category: bottom drains and no rocks.

My reasoning: koi are "dirty" fish. They poop a lot! That's because they eat a lot. Having rocks on the bottom of a pond is just another place for the poop to accumulate and cause problems.

Bottom drains prevent that. They whisk away the poop before it becomes a problem and sends it where it can be separated and broken down efficiently, preventing ammonia spikes.

Rocks are wonderful ... all around the edges. But not on the bottom. So I propose a compromise: When building your pond, lay rocks around the edge, and even under the water line to provide a natural looking edge. But make sure all nooks, crannies, and gaps are filled with black foam. This prevents debris from collecting in places you cannot clean. Don't use rock work any lower than on layer below the waterline — leaving the bottom clean so the debris can be swept away into the drains.

That's the best of both worlds.

Quick-Start Guide: How to Use This Journal

Keeping koi isn't just about water and filters — it's about paying attention to the small changes that, over time, reveal the bigger story of your pond. This journal is designed to make that easy. A few minutes of note-taking each week can save you hours of worry later.

Step 1: Record Water Tests Weekly

At least once a week, write down your test results for:

- pH
- KH (carbonate hardness)
- Ammonia, Nitrite, Nitrate
- Water temperature

These values are the heartbeat of your pond. Over time, you'll see trends: where things usually sit, and when something is drifting out of balance. Catching those shifts early helps prevent disasters.

Step 2: Track Feeding and Treatments

Use the daily log sections to jot down:

- How much and what type of food you gave.
- Any medications, salt treatments, or additives.
- Pond maintenance tasks (filter cleaning, water changes).

This makes it easy to connect changes in your koi's behavior or water chemistry to what you did that week.

Step 3: Observe Your Fish

Your koi will often tell you when something's wrong — if you know what "normal" looks like. Note:

- Swimming behavior (active, sluggish, flashing, gasping).
- Appetite.

- Any visible issues (sores, fin problems, fungus).

Patterns often reveal themselves when you look back through your notes.

Step 4: Seasonal Tasks

At the start of each season, use the space provided for checklists:

- Spring: startup, filter checks, net removal.
- Summer: algae control, feeding patterns.
- Fall: leaf netting, cooling water prep.
- Winter: ice cover checks, aeration, pond heater.

These reminders will help keep your pond in rhythm year after year.

Step 5: Learn from the Logs

Don't worry if you miss a week — just pick up again. The power of this journal isn't perfection, it's accumulation. Looking back over months or years, you'll notice cycles you'd otherwise forget.

Sample Test Parameters Page

Sample Log Entry – KH Tracking

Date	Ammonia NH3	Nitrite NO2	Nitrate NO3	Acidity pH	Hardness KH (drops)
May 15	0	0	20	7.8	3

KH ~53 ppm (low) KH dropped this week. <u>Added 1 cup baking soda</u> to bring KH up toward 100 ppm.

Fed wheat germ pellets. 3 handfuls.

Koi active and feeding well.

Follow-Up: Retest KH tomorrow to confirm rise. Watch for pH stability.

Why KH Matters

KH (carbonate hardness) measures the buffering capacity of your pond water. In simple terms, it's what keeps your pH stable.

Low KH → pH can swing suddenly, stressing or even killing koi.

Healthy KH range → 100–150 ppm (6–8 drops in the test kit).

Raising KH → Baking soda (sodium bicarbonate) is safe, reliable, and inexpensive.

By tracking KH regularly, you'll see drops before they become dangerous, and you can adjust in time to protect your fish.

Printable file:

Water Quality Log

Date	Ammonia NH3	Nitrite NO2	Nitrate NO3	Acidity pH	Hardness KH

Date	Ammonia NH3	Nitrite NO2	Nitrate NO3	Acidity pH	Hardness KH

Illness Journal

If one of your fish becomes ill, make a note of it for future reference. Record the symptoms, treatments, and results — this information can be invaluable if a similar problem occurs again. If you use the downloadable form, you can also attach a digital photo of the symptoms. Over time, these entries will form your own Medical Solutions Record, tailored to your pond.

Fish		Date	
Diagnosis			
Quarantine			
Medicine			
Outcome			

Printable File:

My Koi Collection

Draw your fish! The best way to keep track of your individual fish is to have a visual record. Over time, you'll be able to see how your koi's colours change and mature.

Name Length Variety Age Date Acquired	

Printable Sheet

Here's a quick reference of some of the most common varieties.

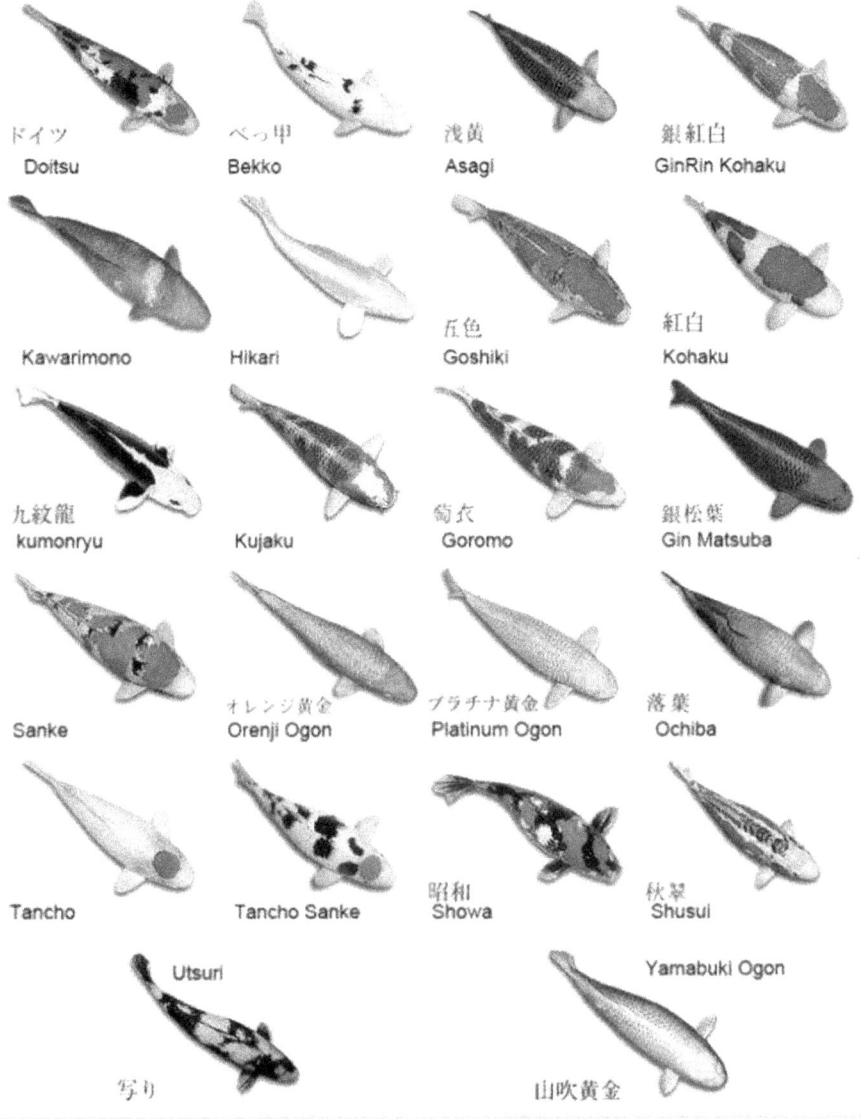

ドイツ Doitsu	べっ甲 Bekko	浅黄 Asagi	銀紅白 GinRin Kohaku
Kawarimono	Hikari	五色 Goshiki	紅白 Kohaku
九紋龍 kumonryu	Kujaku	菊衣 Goromo	銀松葉 Gin Matsuba
Sanke	オレンジ黄金 Orenji Ogon	プラチナ黄金 Platinum Ogon	落葉 Ochiba
Tancho	Tancho Sanke	昭和 Showa	秋翠 Shusui
Utsuri 写り		Yamabuki Ogon 山吹黄金	

Equipment List

Whether you build your own filter or invest in professional equipment, it's essential to keep a record of the details. This information will help you troubleshoot, plan upgrades, and even assist a future pond owner.

Be sure to note:

- Filter type and model
- Manufacturer and serial number
- Installation date
- Media type (and when it was last changed, if ever)
- Where you purchased the filter and media

If you're a DIYer, include:

- Materials used
- Where you sourced them
- Any modifications or design notes

Keeping these details in one place means you'll never have to hunt for receipts or wonder when you last replaced something.

Mechanical Filter

The mechanical filter removes solids from the water before the biological filter.

Biological Filtration
Bioballs provide surface area for bacteria to colonize providing biological filtration

Chemical Filtration
Activated carbon removes organic and inorganic compounds providing chemical filtration

Mechanical Filtration
Filter pads trap waste & debris, providing mechanical filtration

Biological Filter

The biological filter is what "cleans" the water. Clean water for a pond means that there is no ammonia or nitrite in the water that can kill the fish. The process of cleaning the water is known as nitrification. There are two steps: first the ammonia is converted into nitrite, then the nitrite is converted into nitrate. Nitrate (NO3) is relatively harmless to fish unless it is in high concentrations in your pond.

The biofilter uses *bacteria* to do its magic. The bacterial lives in the media in the filter in a brown sludge that collects between and inside the medium of your filter.

DO NOT CLEAN OUT THIS SLUDGE unless there is so much it impedes the flow of water. If this is the case, you need to reevaluate the capacity of the filter/pump for the number of fish you have in your pond.

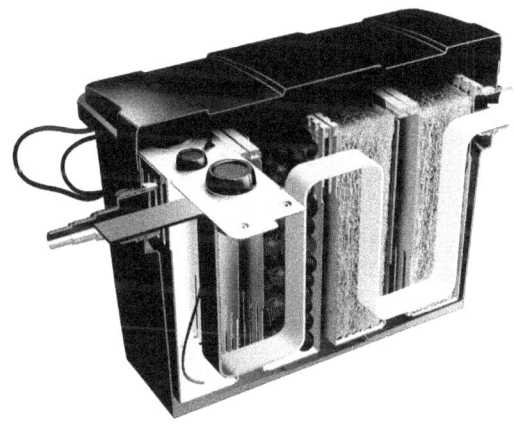

Ultraviolet Filter

The UV filter inhibits the wide-spread growth of free floating, single cell algae. It is not a sterilization light. It won't kill bad bacteria.

A UV light only works if you have the flow rate correct. If you try to pass more water through the light than it can handle, it won't do anything. UV light requires time to kill the algae, so generally, the flow rates are slow. This is impractical for large ponds (>3,000 gallons) because it will take too long to go through all the water in your pond unless you invest in a commercial grade UV light. You are better shading the pond 80% of the day instead.

Waterfall & Skimmer System

Many people use these two devices together, so we will document them together. It is not necessary to have your skimmer plumbed to a waterfall, it can go to a filter and return to the pond a different way. Conversely, your filter can return to the pond by the waterfall to add noise and movement to the water.

It is important to have a skimmer on your pond, to remove debris from the surface, like leaves and floating feces. These would clog your pump or filter if not removed.

Skimmer Unit

A skimmer is necessary to clean off debris from the surface, including oils and residue. Determine how large your skimmer should be by using the surface square foot measurement.

A skimmer should be placed in a leeward spot so that the prevailing wind will push the debris toward it.

A skimmer can either feed a filter, or is commonly used, to feed a waterfall unit. If using a waterfall unit, make sure the skimmer is placed on the opposite side to the waterfall so that it is not drawing in the water that is spilling out of the waterfall, directly.

There are two kinds of skimmers: one is floating, one is installed in the side wall of the pond.

Floating skimmer attached to either a bottom drain or a pump.

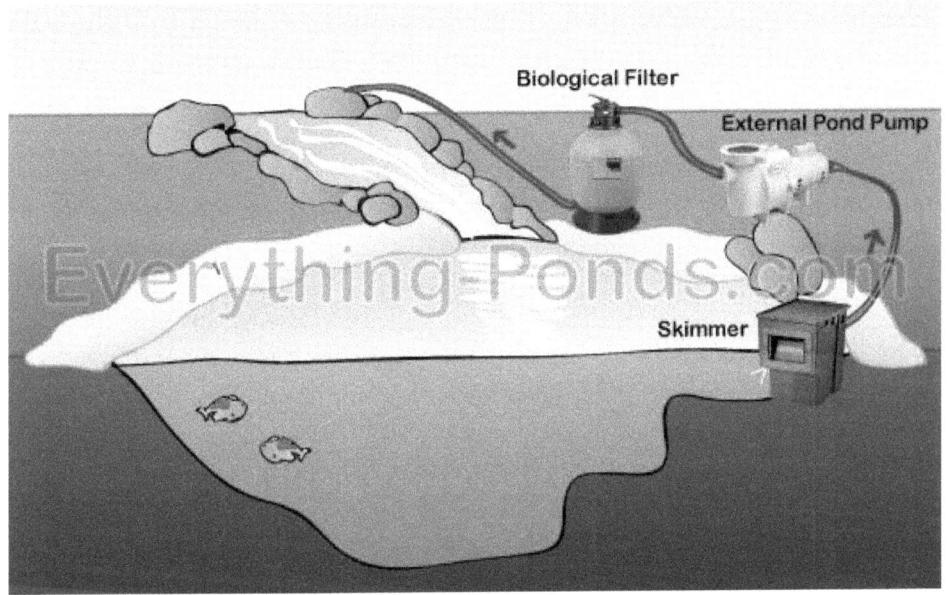

This diagram from Everything-Ponds.com shows a pond skimmer installed in side of pond connected to a filter which outputs to waterfall.

Other Equipment

If you have any other equipment for your pond, for example a reverse osmosis filter, or a carbon filter, this section will let you record the details of each on.

For example, in our pond we use aquarium heaters inside the skimmer tank during winter. This prevents ice from expanding and cracking the skimmer while it's shut down.

In regions where the ground freezes, surface systems such as skimmers and waterfalls should not be operated through the winter. Running them risks ice damage to equipment and can also chill the pond water more than your koi can safely tolerate.

Instead, shut these systems down, protect them from freezing, and rely on aeration or bottom-level circulation to keep water healthy until spring.

Printable worksheet:

Food Diary

Quality food is important to the health and growth of your Koi.

Remember, your Koi will likely grow to be at least 24" long — regardless of the size of your pond. This is why it's important to plan the construction of your pond around that fact, rather than the size you get that at.

What food you end up buying is your choice. But keep track of it, and make sure it's sealed and kept in a dark, cool location. Koi need vitamin C, so make sure whatever you buy has that included.

Some fish eat very quickly, some fish are lazy and eat slowly. Generally, if they know there is never a shortage of food, they will eat slowly, browsing as they need rather than frantically eating in a frenzy.

The rule of thumb is to give no more food that what can be eaten in 10-15 minutes. Measure the food out at first, and watch. Adjust as necessary.

Supplement the regular food with fresh food like thawed shrimp, watermelon, and a head of lettuce. Frozen peas are good too, and some fish like citrus.

We personally use Hikari koi food. This isn't an endorsement, and we're not paid to say it — we simply like the quality for the price. In our experience, it's a good value: the koi thrive on it, and it balances cost with reliable nutrition.

Food Diary

Quantity	Time

Quantity	Time

Quantity	Time

Printable Sheet:

Growth Chart

Koi tend to keep growing at least five years. It's not uncommon for an eight year old koi to put on bulk, or even get larger.

Tracking their growth will help identify problems because there are general growth guidelines available online.

Length of fish in inches or cm. Koi are usually full grown by their fifth year, but keep track of their progress.

Choosing Koi

Not all koi grow to be monsters. Size is largely determined by genetics. The closer a fish's bloodline is to Japanese-bred koi, the larger and more consistent it will grow. Many domestic breeders begin with Japanese stock and produce excellent fish. If size and quality are important to you, stick with reputable koi breeders.

If you plan to show your koi, bloodline information is essential. Serious breeders provide pedigree details that pet stores usually cannot.

Pet stores often carry koi alongside goldfish, but unless the store has a koi enthusiast as manager, their stock is usually made up of culls from breeders — the "rejects." These fish may have irregular patterns, uneven growth, or shorter lifespans. Some will surprise you and grow into big, beautiful pond fish, while others remain small and plain. Buying from a pet store is a gamble — though sometimes it pays off wonderfully.

Koi are often grouped into three broad categories:

1. Championship quality — show fish with strong bloodlines.
2. Serious hobbyist quality — well-bred koi, often from Japanese or domestic breeders.

3.   Pond/garden quality — the everyday koi most of us keep and love, regardless of markings.

Most koi keepers fall into the third category — and that's perfectly fine. These fish may never win a trophy, but they bring beauty, movement, and joy to the pond, which is what matters most.

 Informative Video on YouTube: "How do I Choose a Good Koi Fish?" by The Koi Partner.

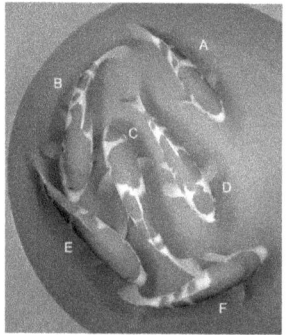

High quality Koi   Good quality Koi   Pond quality Koi

Printable Sheet:

Maintenance & Procedures

Many seasonal tasks in koi keeping are repeated year after year: cleaning the skimmer, maintaining the filters, refreshing the water. Writing down exactly what you do and how often you do it is invaluable. It helps you stay consistent, but it also creates a record that can be passed along if someone else takes over your pond one day.

Clear documentation is also a great tool for anyone looking after your pond while you're away. A house sitter, neighbor, or friend can follow your notes step by step without guesswork.

For this reason, don't just log the big seasonal jobs — record your daily procedures as well. Feeding routines, observation habits, even how you check for debris or surface bubbles are all part of your pond's rhythm. The more you write down, the easier it is for you — and for anyone helping you — to keep your koi healthy and happy.

Spring Procedure (sample)

Procedure: _____SPRING START UP_____

Frequency: _____Spring (April) When the Water Is Free of Ice_____

Process:

1. Remove pump from skimmer, clean thoroughly, check wire for breaks and plug for defects or wear.
2. Drop water level below skimmer intake, and drain skimmer bucket, ensuring pipes are clear.
3. Check waterfall filter, remove debris, old leaves, plants, rocks, etc that could be hindering water flow.
4. Flush out waterfall tank and sponge.
5. Flush and drain filter to remove sludge and toxic substances.
6. Check all connections on filter, do a pressure test to ensure pressure vessel is still in good working order. Test valve works at all positions.

7.   Reinstall pump in skimmer, ensure it's connections are tight and secure.

8.   Restart pump and waterfall. Watch for problems and water pressure.

Mapping Your Plumbing

If you're like me, you'll forget where all the underground connections are about a year after the pond is built. That's not a problem — until there's a leak. Then it becomes a big one.

The best way to prevent that is to make a diagram of your plumbing system. Sketch out:

- Pipe sizes
- Types of connectors
- Where each line runs
- Valves, drains, and any special fittings

If possible, draw it to scale. Even a rough illustration is better than nothing — but the more detail you include, the more useful it will be in the future.

I scribbled my first one by hand, and later redrew it on the computer. Both versions have saved me more than once.

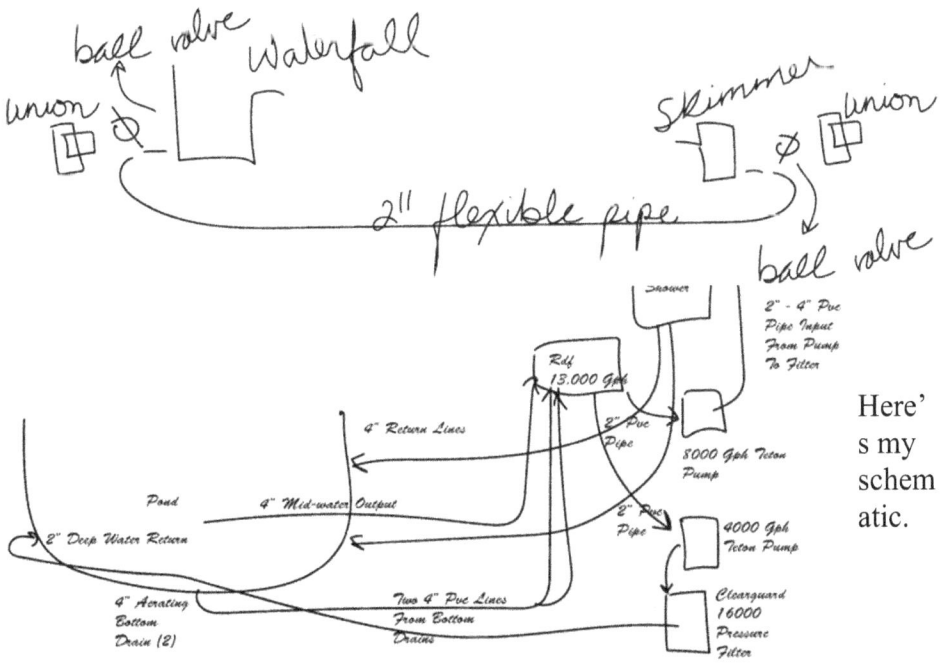

Here's my schematic.

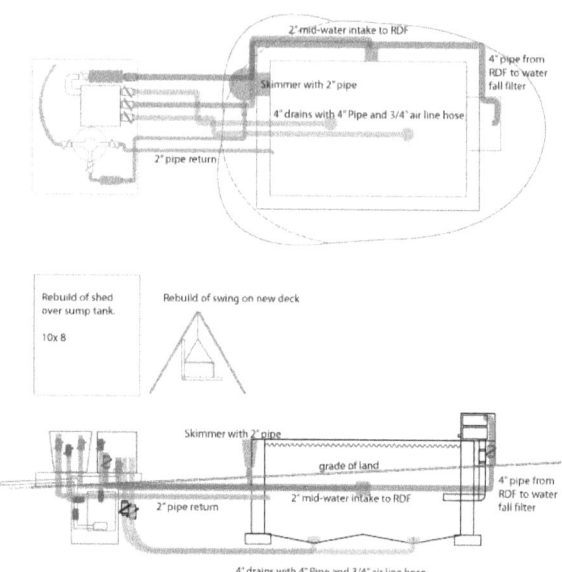

Printable Worksheet:

New Pond Designs

As soon as you've built your first pond, you're already designing the next one.

I know, you think it won't happen to you — but it will. Why? Because technology changes. New ideas appear. Old ideas get refined. Your fish get bigger. And your own interests in the hobby shift. There are countless reasons, but sooner or later, every koi keeper starts imagining the next pond.

When that happens, plan it out. Start with two lists: "What I did wrong" and "What I can do better." Then add the "nice-to-haves" at the end.

Remember, you don't have to make every improvement at once. Many upgrades can be done incrementally. And when you've upgraded everything but the container itself… well, that's when you know it's time for a new pond.

One important note: if you already have fish, you'll need a place to keep them safe while you build. (And yes, it will be a bigger pond.)

Finally, don't just dream — draw. Use graph paper to sketch your ideas. Seeing your next pond on paper is the first step toward making it real.

Medicine Chest

Keep a stock of essential medicines and chemicals on hand. Emergencies rarely give warning, and having supplies ready can mean the difference between saving a fish and losing one.

Some items are useful year-round, such as:

*   Dechlorinator (for water changes)
*   Beneficial bacteria products (to support filter health)

Other items are for special circumstances:

*   Bottled oxygen with an air stone (for reviving a distressed fish)
*   Specific medications (parasite, bacterial, or fungal treatments)

When you purchase medicines, always note their expiry dates. Once expired, their effectiveness is no longer guaranteed — and using them can do more harm than good. Keep your records current so you know what is safe to use.

For example:

Product	Use	Qty on Hand	Expires
Rock Salt	Good for skin and gills	20 lbs	no
Seachem Stability	Adds bacteria to filter	5 liters	no
Seachem Prime	Detoxifies NH3, NO2, and chloramine	1 gallon	no
Praziquantel	treats parasites like flukes and tape worms	4 oz	yes

Product	Use	Qty on Hand	Expires
Proform-C	Treats ich, costia, trichodina, chilodonella, oodinium, and fungus	1 litre	yes
Tincture of Iodine 10%	Abrasions, scrapes, minor cuts		2-3 yr
Clove Oil	Euthanasia		2-3 yr
Potassium Permanganate	disinfectant, oxidizer, bacterial ulcers		no if kept cool, dark
Hydrogen Peroxide	10% neutralizes PP.		1-2 yr
Dimilin	treats anchor worms, fish lice		yes

Printable Worksheet:

List of Koi Vets

You should have the name of a vet who can diagnose your fish in the event that your local pet store cannot.

Sometimes, you'll need a vet to prescribe medications as well.

Dr. Erik Johnson, is a long time Koi Vet and in general, specializes in fish. He has several books available to purchase with lots of good information and procedures for treating your Koi.

Website: https://drjohnson.com/
Twitter/X: https://twitter.com/erikjohnsondvm

Books: Koi Health & Disease, by Erik L. Johnson, DVM
Advanced Koi Care, by NicholasSaint-Erne, DVM

My Vet: _____

Location: _____

Phone number: _____

Email: _____

Website: _____

Books on Amazon

Nicholas Saint-Erne, DVM

 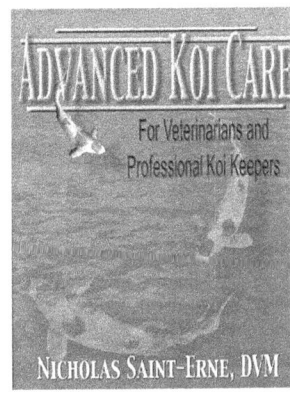

Medical Equipment

Eventually, there will come a time when you'll need to diagnose a health issue with your koi. Unless you have ready access to a fish veterinarian, you'll want to be prepared with the right tools.

The good news is that most koi diseases are well-documented and relatively straightforward to identify once you have the basics.

Here's a starter kit every koi keeper should keep on hand:

☐ Microscope (capable of at least 400x magnification)
☐ Slides and cover slips
☐ Sterile swabs
☐ Spoons or mixing tools
☐ Measuring cups and spoons
☐ Small-gauge needles (insulin syringes work well)
☐ A tote or case to store everything together

With these tools, you'll be able to make accurate observations, perform basic scrapes, and match what you see with established koi health guides. It's not about replacing a vet, but about giving yourself the ability to act quickly and knowledgeably when problems arise.

Some Additions to Your Diagnostic Kit

☐ Cover slips → essential with slides for proper microscope work.
☐ Scalpel or razor blade → for taking skin/gill scrapes (very small, careful use).
☐ Pipettes or droppers → for handling liquids or transferring small water samples.
☐ Latex or nitrile gloves → for hygiene and personal protection.
☐ Small flashlight or headlamp → for examining koi closely at night or in shaded areas.
☐ Magnifying glass / jeweler's loupe → quick checks without a microscope.
☐ Notebook or printed forms → to record findings (pairs beautifully with your journal idea!).

☐ Digital scale (small) → handy for weighing medication dosages for individual fish.

☐ pH/Ammonia test strips → quick checks alongside microscope work (backups to liquid kits).

Optional but Helpful

☐ Tricaine Methanesulfonate (MS-222) or clove oil → if you need to sedate a fish briefly for examination.

☐ Small viewing bowl/tub → to catch and hold fish for inspection under natural light.

REMEMBER: *Never medicate blindly — diagnose first.*

Emergency Preparedness

Emergencies happen — to us, to the fish, and to the property. We never know what life will throw our way, but we can plan for it.

Create a list of emergency contacts: people who know your fish, understand your pond system, and are capable of running it if you cannot. Review this with them at least once a year, so they're familiar with your setup and procedures.

Have an emergency procedure in place, just as you would for a tornado or hurricane drill. Know exactly what to do before you ever have to do it. And practice — because in a real crisis, clear steps matter more than good intentions.

Emergency Contacts should include:

1. Someone qualified to care for your koi in your absence.
2. Someone who can help in the event of a major weather or property emergency.
3. Backup contacts in case your first choice is unavailable.

Keep this list printed and posted near the pond or filter system, where it's easy to find.

Printable Worksheet:

Planning for the Future of Your Koi

It's not a pleasant subject, but it's one every koi keeper should face. Koi can live 25 years in captivity — sometimes much longer with the right systems in place. They are long-lived companions, and if you die before them, you need a plan so your executor knows what to do.

Bequeathing Your Koi
Consider naming a beneficiary for your koi, just as you would for other valuable assets. Options include:

- A public space that already keeps koi, such as a commercial building, garden, or zoo.
- A trusted member of your local koi community or club.
- A reputable dealer or pet store that has agreed, in writing, to accept them.

If leaving your koi to an individual, include a stipend for care. Even experienced hobbyists can be overwhelmed by the sudden expense of taking on large fish.

Leaving Koi with the Property
Another option is to leave the koi with the pond on your property. If your executor is selling the house, or if the property passes to someone in your will, make sure they understand the responsibility of caring for the fish — and that they agree to it.

Put It in Writing
Whichever option you choose, discuss it with a lawyer and make sure your wishes are written clearly into your Last Will and Testament. This provides peace of mind for you, your family, and — most importantly — for the fish you've cared for.

Appendix A- Pipe Sizes & Flow Rates

For PVC pipe and flex PVC pipe—Schedule 40

Pipe Size	Inside Diameter	Typical Flow (GPH)	Notes
1 ½"	1.61"	~3.500	Small ponds, short runs only
2"	2.07"	~4,800-5000 GPH	Standard for most ponds
3"	3.07"	~9,000 GPH	Good for gravity fed RDFs
4"	4.03"	~13,000 GPH	Bottom drains and main lines

NOTE Larger pipes reduce friction loss and improve efficiency.

Appendix B - Pipe Types

Pipe Type	Strength	Pressure Rating	Best Use	Notes
ABS	Lightweight, brittle, black	Not pressure-rated	Gravity-fed lines	Easy to cut/glue; cheaper, but not for pumps.
PVC (Sched 20)	Thin-walled, white	Light duty	Central Vacuums	Not for water use
PVC (Sched 40)	Strong, thick, white	Pressure-rated 120-160 PSI	Pump lines, buried pipe	Gold standard for ponds
PVC (Sched 80)	Extra thick, grey	Heavy duty 200-280 PSI	Industrial	Overkill for ponds.
PVC Conduit	Flexible, grey	No	Electrical, running wires	Not for Water
Flex PVC 2" or 4"	Black or White, Strong	Pressure rated ~100 PSI	Long runs – allows for curves without elbows	Not good for tight spaces Not UV resistant

Appendix C - Valve Types

For PVC Pipe and Flex PVC systems

Type	Use	Pros	Cons	Notes
Ball Valve	On/off flow control	Cheap, reliable, simple	Hard to adjust flow precisely	Use full-port so the inside opening matches the pipe size
Gate Valve	Fine flow control	Very precise adjustment	Large, bulky, more expensive	Great for gravity-fed systems
Check Valve	Stops backflow	Keeps pumps primed	Can stick open or closed	Install vertically or on horizontal runs with swing type
Union Ball Valve	Isolation & disconnection	Easy equipment removal	Slightly larger & pricier	Best before and after pumps, filters, UVs
Swing Check Valve	Stops reverse siphon	Silent, low resistance	Needs horizontal run	Perfect after external pumps
Spring Check Valve	Stops backflow in any position	Works vertical or horizontal	Adds resistance and can clog	Only use on clean water lines

Pro Tip: Always place valves and unions before and after major equipment (pumps, filters, UV, heaters) so you can remove them easily. Use true-union ball valves wherever possible—they last longer and make servicing painless. Avoid too many valves: every one adds friction loss.

Appendix D - Glue vs Cement

First of all, you don't glue PVC pipe together. You weld it with a two step process. It's called PVC Cement.

What's the Difference?
PVC cement doesn't just stick pieces together—it chemically melts the surfaces and fuses them into one solid piece.
PVC glue (like wood or craft glue) is not for pressure plumbing and will fail under water pressure.
So when working with PVC pipe or flex PVC pipe for pond plumbing, always use PVC cement—never ordinary glue.

Primer + Cement: The Right Method
- Use PVC primer (purple or clear) first—it softens and cleans the surface.
- Apply a generous, even coat of PVC cement to both the pipe and the fitting.
- Push the pipe fully into the fitting and twist ¼ turn to spread the cement evenly.
- Hold for 30 seconds while it sets (especially on large pipe).
- Let joints cure at least 24 hours before pressure-testing.

NOTE For flex PVC, use medium- or heavy-bodied PVC cement (often labeled "for flexible PVC") so it bonds through the soft wall fully.

Common Mistakes
- Skipping primer → weak, leaky joints
- Using glue instead of cement → joint failure under pressure
- Not pushing pipe in fully → dry spots → leaks
- Moving the joint before it sets → breaks the seal

Appendix E - Weatherproofing

Protecting your pipes from frost, floods, and storms

Frost Protection
- In cold climates, bury your PVC pipe or flex PVC pipe below the local frost line (depth where soil freezes solid).
- Frost line varies:
- ~12" (30 cm) in mild zones
- 36–48" (90–120 cm) or more in cold northern zones
- Pipes **must slope to drain** if they'll be emptied seasonally.
- Where burying isn't possible, **insulate pipes** with foam wrap and use **pipe heating cable**.

NOTE   Frost heave can shear fittings right off. Keep all buried plumbing **below freeze depth and on undisturbed soil**.

Flood Resilience
- Place pumps, electrical, and filters above known flood level (or on raised platforms).
- Install check valve on return lines to prevent reverse siphon when water rises.
- Anchor filter housings and pipes to resist floating or shifting.
- If you're in a high water table area, weigh down buried pipes with sand backfill or strapping.

NOTE   Flood water is full of silt—design your pond to keep it **out of the pond system**, not just off your lawn.

Storm & Wind Protection
- Use flexible couplings near equipment to absorb vibration and ground movement.
- Anchor exposed pipes, pumps, and filters so high winds can't shift or topple them.
- If hurricanes or severe windstorms are possible:
- Secure covers and domes
- Drain or lower water level slightly to allow for heavy rainfall

- Disconnect and cap above-ground plumbing if debris impact is likely

NOTE In extreme events, **protect the infrastructure first**—fish can survive without pumps for a while, but pumps won't survive flying tree branches.

For More Information Online on Frost Line Reference Links

Wikipedia Climate
Zones for America

Canadian Climate
Hardiness Zones

NOAA

Appendix F - Conversion Charts

Temperature Reference for Koi Health

Condition	°F	°C	Notes
Fish immune system active	50 °F	10 °C	Below this, immune response is sluggish
Fish best temperature range	75 °F	24 °C	Optimal growth & activity
Fish too hot (stress zone)	80 °F	27 °C	Oxygen stress, higher risk of disease
Coldest limit (survivable)	46 °F (approx)	8 °C	Koi survive but metabolism slows
Stop feeding protein, switch to wheatgerm	59 °F	15 °C	Easier digestion at cooler temps
Stop feeding entirely	50 °F	10 °C	Digestion shuts down
Must use aeration	78 °F+	26 °C+	Warm water holds less oxygen
"Aeromonas Alley" (disease risk zone)	up to 50 °F	up to 10 °C	Bacteria active before immune system is

Volume, Weight & Pump Sizing Reference

Pond Size	US Gallons	Liters	Weight of Water*	Minimum Pump Size**
Small	500 gal	1,893 L	4,170 lbs / 1,890 kg	500 gph / 1,900 L/h
Medium	2,000 gal	7,570 L	16,680 lbs / 7,570 kg	2,000 gph / 7,600 L/h
Large	5,000+ gal	18,925+ L	41,700+ lbs / 18,925+ kg	5,000 gph / 19,000 L/h

* Water weighs 8.34 lbs per US gallon (1 kg per liter).

**Pump rule of thumb: turn over the entire pond volume at least once per hour. Bigger is better — especially with koi.

Stocking Rule of Thumb
- 1 adult koi = 500 gal (1,900 L) minimum.
- 500 gal pond → 1 koi max.
- 2,000 gal pond → 4 koi max.
- 5,000 gal pond → 10+ koi possible, depending on filtration & aeration.

Salt Treatment Reference

Purpose	% Salt	Metric Dose	Imperial Dose
Tonic / stress relief	0.1–0.3%	1–3 kg per 1000 L	0.83–2.5 lb per 264 gal
Nitrite spike protection	0.6%	6 kg per 1000 L	5 lb per 264 gal
Parasite treatment (short bath)	1.0–1.5%	10–15 kg per 1000 L	8.3–12.5 lb per 264 gal

Tips:
- Always dissolve before adding, or place in a waterfall to dissolve gradually.
- Salt stays until removed with water changes.
- Never combine salt with formalin or potassium permanganate.

General Conversions

Volume
1 US gallon = 3.785 liters
1 liter = 0.264 US gallons
1 cubic foot = 7.48 US gallons = 28.3 liters

Weight
1 pound = 0.454 kilograms
1 kilogram = 2.2 pounds

Salt / Additives
1 kg salt per 1,000 L = 0.83 lb per 264 US gallons
10 g per 100 L = 0.083 lb per 26 US gallons

Temperature
$(°C × 1.8) + 32 = °F (°F − 32) ÷ 1.8 = °C$

Quick Pond Rule of Thumb
1,000 gallons = 8,340 lbs (3,785 liters = 3,765 kg)
1 acre-foot of water = 325,851 gallons = 1,233 cubic meters

Resources & Downloadable Files

Please find the QR codes for all the downloadable forms I have created for this book. It includes access to your very own Pond Calculator file that you can copy.

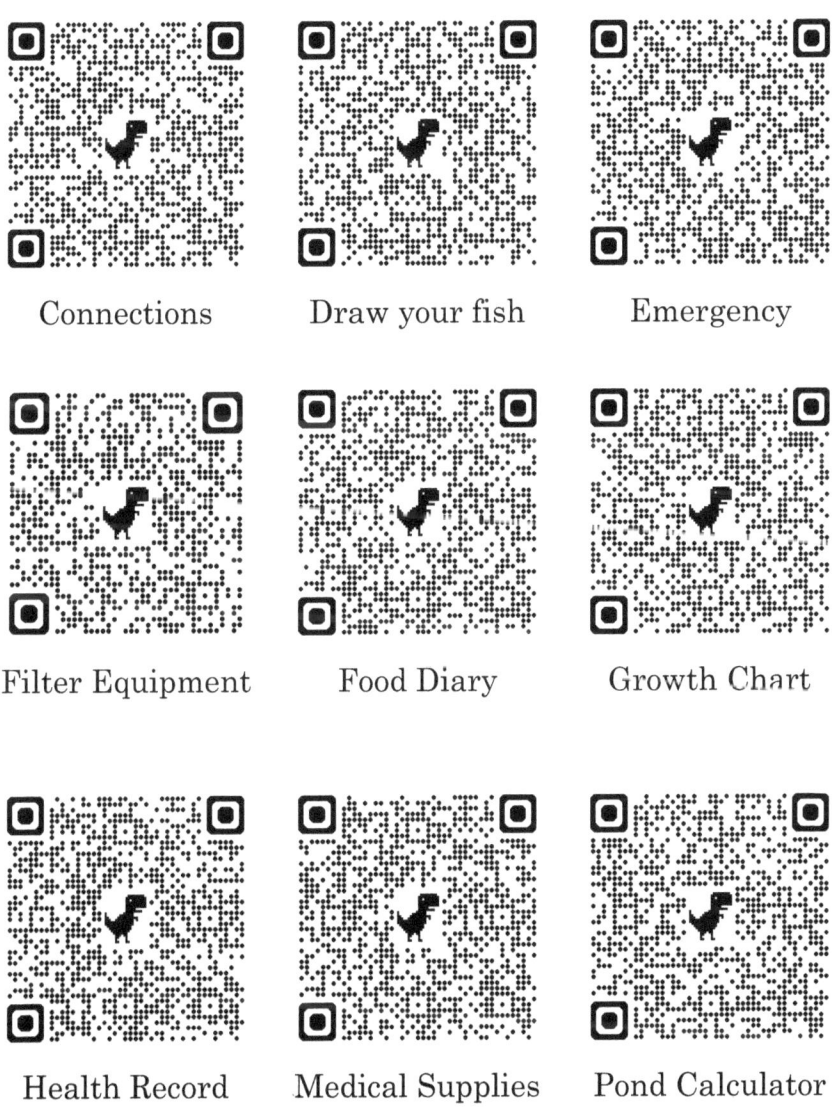

Connections	Draw your fish	Emergency
Filter Equipment	Food Diary	Growth Chart
Health Record	Medical Supplies	Pond Calculator

Regular   Water Quality Log
Procedures

Index of Terms

www.ingramcontent.com/pod-product-compliance
Lightning Source LLC
Chambersburg PA
CBHW051334120626
46547CB00016B/2542